Data, Chance & Probability

Grades 4-6
Activity Book

by Graham A. Jones
& Carol A. Thornton

© 1993 by Learning Resources, Inc.
Lincolnshire, Illinois 60069

ISBN: 1-56911-997-X

Printed in the United States of America.

DATA, CHANCE AND PROBABILITY gr 4-6

Table of Contents

Introduction..4

Distances

Teaching Notes.....................................7
How Far? 8
How Many? 9
Graph It! 10

Baseball Cards

Teaching Notes....................................11
Free Baseball Card 13
3-Card Special.....................................14
NL First...15
Why NL?..16
Match the Cards..................................17
Cards and Cubes..................................18
Rate a Team 19
Which Card? 20
AL for Sure 21
Same Chance 22
Why the Same? 23
What Chance AL? 24
Why AL?...25
Card Match...26
Change Chances...................................27

Get the Message

Teaching Notes....................................28
How Many Shows? 30
Box the TV Data 31
Plot the TV Data 32
Make a Box-and-Whisker Plot 33
TV Time 34
TV Ratings 35

Candy Sort

Teaching Notes....................................36
Take a Candy 38
Candy Chances 39
Match the Candies 40
Mini Candies 41
Double Red 42
Mini Packs 43
More Mini Packs...................................44
Better Chances.....................................45
Greater Chance....................................46

Pizza Toss

Teaching Notes 47
Kinds of Pizza 49
Pizza Pieces 50
Box the Pizza 51
Pizza Favorites 52
Topping Survey......................................53
Pizza Data 54
Pizza Show 55

Out of the Bank

Teaching Notes 56
In the Bank 57
Just as Many 58
3 Coins 59
Even It Up..60
Nickel Up 61
Two Come Out 62
A Chance for Two..................................63
Match the Coins 64

Game Time

Teaching Notes......................................65
Who Goes First? 67
Now Who?..68
Are the Chips Fair? 69
Is It Fair? 70

Pancake Fun

Teaching Notes 71
Blueberry Pancakes...............................72
No Blueberries 73
Enough Blueberries 74

Activity Master 1.....................................75

Activity Master 2.....................................76

Family Gram 77

Award Certificate 78

Good Work Award....................................79

Progress Chart 80

Introduction

This *Data, Chance, and Probability Activity Book, 4-6* contains 55 problem-solving activities and ideas for students in grades 4-6. Using real world situations and data from cross-curricular topics, students explore probability and data analysis. Learning is accomplished by mathematical modeling and integrating probability with data analysis when possible.

The problem-solving activities are presented in cooperative, active, learning settings. Students:

- **Explore a problem**
- **Predict outcomes**
- **Gather data and other information**
- **Organize the information**
- **"Mathematize" the situation (assign numbers or measurements)**
- **Communicate results**
- **Reflect on and extend ideas**

As a carefully structured supplement to your mathematics program, *Data, Chance, and Probability Activity Book, 4-6* enhances the development of mathematical ideas and furthers students' understanding of probability and data analysis. Activity pages emphasize the following:

- **Outcomes and events**
- **Data analysis and presentation**
- **Probability comparisons**
- **Numerical probabilities**
- **Arrangements**
- **Mathematical modeling**

Data, Chance, and Probability Activity Book, 4-6 follows recommendations in NCTM's *Curriculum and Evaluation Standards for School Mathematics* (1989). Activities focus on four major goals of this NCTM document: problem solving, reasoning, communication, and making connections. Activities also reflect the spirit of the NCTM 5-8 Addenda Series on *Dealing with Data and Chance. Data, Chance, and Probability Activity Book, 4-6* is a useful supplement to existing mathematics programs in regular, special education, and remedial classroom settings.

Each activity page presents a hands-on learning experience for the students. The three parts of each activity provide opportunities for students to:

1. Address a problem.

2. Communicate results.

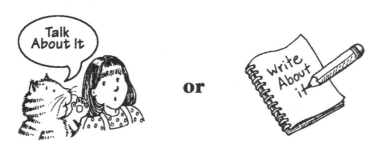

3. Interpret and extend ideas.

To encourage reflective thinking and increase constructive learning, students follow a two-step approach. First, they address a problem and predict the results. Then they experiment with a partner, collect and quantify data, and discuss the results. This two-step method is extended when students review their thinking with partners, then share their findings and interpretations with the class. This sharing and discussion helps students find new applications for the ideas presented.

Using the Activities

The *Teaching Notes* summarize the content of the activities and offer suggestions for guiding the work on the activity pages. Most activities require students to work with partners. In the "Explore" part of the activities, students collect and present data, first with their partners and then as a class.

Students are encouraged to "Write About It" or "Talk About It" as a way of communicating their findings. Finally, in the "Think and Tell" section of the activities, students are challenged to quantify, interpret, apply, or extend the ideas discussed.

Teaching Notes
Distances

Materials Needed:

• 2" x 2" Paper Squares in Two Colors •
Chart Paper

Warm-Up

To introduce this section, review how to read a graph by presenting one from a magazine, newspaper, or textbook. Have students talk about and interpret the graph Then discuss ways of traveling to and from school — by foot, bicycle, school bus, car, or public transportation. Help students identify how they travel to and from school, and make a list on the chalkboard.

Using the Pages

How Far? *(page 8)*: Establish guidelines for determining distances, and set the unit of measure. An easy way to find the median is to have students line up in order, from the closest to the farthest distance from school. Then, in the case of an odd number of students, select the middle student as the median. In the case of an even number of students, select the average of the two in the middle.

How Many? *(page 9)*: Review the terms in the Word Box, and help students compile the data in the chart. Discuss possible topics that can be used to make other frequency tables.

Graph It! *(page 10)*: Encourage students to construct the double bar graph. Explain that the graph presents information in a way that allows us to compare groups more easily.

Wrap-Up

Have the class consider other distances they can use to compile data, then have them graph the information. Use distances from each student's house to a shopping center, to the post office, or other community locations.

How Far?

Name_____

 **xplore** Find out how far you live from school.

Use blocks, miles, or any standard measurement.

Record the distance on a paper square.

Data Activity

- Work together as a group.

- Line up in order from those who live closest to those who live farthest away.

- Those who live the same distance stand next to each other.

- Record the information on a chart as shown.

How Far?

Distance in Blocks					
1	2	3	4	5	6
face	face	face	face	face	face
face	face	face	face	face	
face		face	face		face

Talk about the information:

- What is the range?
- What is the median?

What other ways can you show the data?

Need: 2" x 2" paper squares, in two colors (one for boys, one for girls); chart paper.

Save: How Far? chart for How Many? (page 9) and paper squares for Graph It! (page 10).

How Many?

Name_____

 **Explore** Using the information from your *How Far?* chart:

- Find where the distances cluster.
- Find the mode.

Data Activity

- Make a frequency table and name it *How Often?*
- Use the information from your *How Far?* chart.

How Often?

Blocks	Tally	Frequency
1	II	2
2	I	1
3	IIII	4
4	III	3
5	卌 II	7
6	II	2
7	卌	5

 **Talk About It**

Where do the distances cluster?

What is the mode?

Word Box

Cluster:
place where many items occur

Mode:
number that occurs most frequently

 **Think and Tell**

Can you make a frequency table for other data?

Need: *How Far?* chart from page 8.

Data, Chance, and Probability Activity Book, 4-6
© 1993 Learning Resources, Inc.

Graph It!

Name_____

 **xplore**   Divide the class into two groups using the color squares from *How Far?* (page 8).

Find the range for each group.

Find the median and mode for each group.

Graph Activity

• Use the color squares to make the double bar graph named *How Far?* for Group 1 and Group 2.

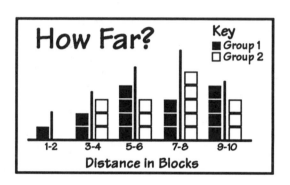

How Far?

Key
■ Group 1
□ Group 2

1-2 3-4 5-6 7-8 9-10
Distance in Blocks

Talk About It

For each group:
- Find the range.
- Find the mode.
- Find the median.

Think and Tell

How do the two groups compare?

Divide the class into two new groups and graph the data.

Find the range, median, and mode for each group.

Need: Paper squares from *How Far?* (page 8).

Data, Chance, and Probability Activity Book, 4-6
© 1993 Learning Resources, Inc.

Teaching Notes
Baseball Cards

Materials Needed:

Red/White Counters (chips) • Paper Plates •
• Overhead Transparency • Number Cubes •
• Red & White Gummed Dots • Mural
Paper • Paper Squares • Activity Master 2

 Warm-Up

These activities lead to the assignment of numerical probabilities. Concentrate first on certain and impossible events, as these concepts start students thinking about probability, namely 1 (certain) and 0 (impossible). Encourage students to identify "certain" and "impossible" events throughout the school day, and assign 1 or 0 to each event accordingly. As an extension, have students identify some possible (but not certain) events, and predict their probabilities between the two extremes of 1 and 0.

Introduce the lesson in this section with a discussion of baseball or other sport of interest to your students. Help students identify the two major U.S. leagues — AL (American League) and NL (National League). Ask volunteers to provide information about baseball, and explore baseball card collecting.

 Using the Pages

Free Baseball Card *(Page 13)*: The purpose of this page is to help students realize there are two possibilities: AL and NL. However, the pattern of these outcomes cannot be predicted with certainty. Remind students that a prediction is a *guess* based on the information given.

3-Card Special *(Page 14)*: As students work in pairs, one student should set up the experiment. The chip outside should be a white chip; chips under the plate should be red. Point out that in an experiment, an outcome is a possible result. If an outcome can happen, it is a *possibility*. If only one outcome is possible, it is a *certainty*. Encourage students to use words like *outcome, certain,* and *possible* in context.

NL First *(Page 15)*: Some students may choose all red chips, others all white, and others a combination. Variable results occur with limited samples.

Why NL? *(Page 16)*: In constructing the class graph, total the data from *NL First*. After the results of three groups have been recorded, find the ratio of NL to AL. As the results are combined, the outcomes should approach a ratio of 2 to 1.

Match the Cards *(Page 17)*: Help students conclude there are two possible answers. Students might give reasons why both 2/3 and 4/6 model the situation. In the "Think and Tell" section, students can respond correctly by keeping the shaded area in a 2 to 1 ratio (8 to 4, 16 to 8, or even a total of 240° to 120°).

Cards and Cubes *(page 18)*: For the "Think and Tell" section, students should discuss in numerical terms why the number cube, the spinner, and the card models give the same answer to the problem. Use this activity to make connections between ratios and fractions.

Rate a Team *(page 19)*: Help students make the floor line plot. Then review the terms *range*, *mode*, and *median*. Encourage students to work independently to calculate each of these figures for the data.

Which Card? *(page 20)*: Explain that a tree diagram helps find all possible outcomes. Review the diagram and help students follow the routes to determine all possible cards.

AL for Sure? *(page 21)*: After focusing on "certain" and "impossible" events, students assign numerical probabilities of 1 and 0, respectively.

Same Chance *(page 22)*: Be sure students put the chip back before selecting another chip; otherwise, results will be skewed from 50/50.

Why the Same? *(page 23)*: To compare the ratio of AL to NL, students should tally in blocks of five. Students can then determine the chances of getting AL and NL. Although unlikely events can occur, the class data should produce probabilities that approximate 1/2.

What Chance AL? and **Why AL?** *(pages 24 and 25)*: These activities are similar to *Same Chance* and *Why the Same?* activities. Again, the probabilities may only approximate 2/8 for NL and 6/8 for AL.

Card Match *(Page 26)*: In this activity, students make a connection between equivalent fractions like 3/4 and 6/8. In the "Think and Tell" section, they find other equivalent fractions such as 9/12, 12/16, or 15/20.

Change Chances *(Page 27)*: This problem-solving activity requires students to determine the correct kind and number of cards to put in the box (sample space) for a particular outcome. There are many possible answers, all requiring the same number for each league (e.g., 1 NL, 1 AL; 2 NL, 2 AL).

 Wrap-Up

In the *Change Chances* activity, NL was twice as likely to occur as AL. Challenge students to identify other situations where one event is twice as likely as another. Then ask students to decide how, if possible, to make the events equally likely.

Free Baseball Card

Name_____

 **E**xplore

1 Free Baseball Card

AL
or
NL

Which baseball league might be in the box?

If you get one league in the box, will you always get the other league in the next box?

Your Turn

The Experiment

Find a partner.

Experiment with a red/white chip:

- white side for AL
- red side for NL

The Experiment

- Close your eyes and turn over a chip several times in your hand. Predict the color.
- Open your eyes and record the color.
- Do it again.
- Partner takes a turn.

 *Talk About It*

Did the colors turn out as you predicted?

How do the results compare with those of other classmates?

Your Turn

White	Red

Partner's Turn

White	Red

 *Think and Tell*

How do the chips model the baseball cards?

Word Box

Predict:
Make a guess based on information

Model:
Give the same results

Need: 1 red/white chip; Activity Master 2

3-Card Special

Name_____

 **E**xplore

3 Free Baseball Cards

NL   AL   NL

An AL card is taken out of the box first.

Which league do you think you will find next?

Your Turn

Find a partner.

Experiment with chips:

- 1 white for AL
- 2 red for NL

The Experiment

- Hold the white chip and put the red chips under a plate.
- Draw one chip from under the plate and tell the color.
- Predict which color you'll draw next.
- Check the color, and show your partner.

 **Talk About It**

Did the outcome turn out as you predicted?

Word Box

Outcome: possible result

Certain: definite result

 **Think and Tell**

What was *certain* to happen? Why?

If an NL card had been found first, could you predict the next outcome with certainty? Explain.

Need: 2 red chips, 1 white chip; paper plate; Activity Master 2

 (14)

NL First

Name_____

xplore

3 Free Baseball Cards

NL   AL   NL

Which league do you think you will find first?

Your Turn

Experiment with chips:

- 1 white for AL
- 2 red for NL

The Experiment

- Put 3 chips under a plate.
- Draw one chip.
- Record the color on the Tally Sheet and put the chip back.
- Predict the next outcome.
- Repeat 5 times.

Talk About It

Did the outcome turn out as you predicted?

How do your results compare with those of your classmates?

Tally Sheet

Red	White

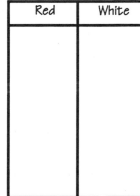
Think and Tell

Which color came up more often?

Why?

Need: 2 red chips, 1 white chip; paper plate; Activity Master 2

Save: Tally Sheet for Why NL? (page 16).

Why NL?

Name_____

 **xplore** Work in small groups.

Combine the Tally Sheets from *NL First* to develop group totals.

Graph Activity

Combine the group totals. Use tally marks to make a class graph of the *NL First* activity data.

Total the tally marks.

- NL_____

- AL_____

Class Graph: NL and AL						
NL (red)	ՊԱՌ	ՊԱՌ	ՊԱՌ	ՊԱՌ	‖	
AL (white)	ՊԱՌ‖					

Talk About It

How does the class graph compare with:

- your own results?
- your group's results?

How do the NL and AL totals in the class graph compare?

Think and Tell

Use the class graph to compare the chances for NL cards to AL cards.

How do these chances compare to the actual number of cards in a box?

Need: Tally Sheet from *NL First* (page 15); transparency for class graph.

Save: Class Graph for *Cards and Cubes* (page 18).

Data, Chance, and Probability Activity Book, 4-6
© 1993 Learning Resources, Inc.

Match the Cards

Name_____

 **E**xplore *NL First* had three free baseball cards — two NL cards and
one AL card.
Which spinners work like the cards in *NL First?*
Label each shaded part *NL* and each unshaded part *AL.*

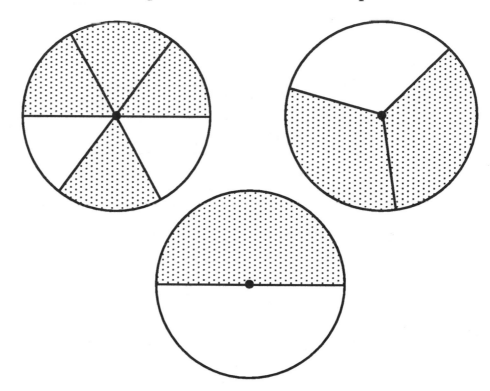

 Tell why the spinners you chose are like the
baseball cards in *NL First.*

 Can you make another spinner
that would work?

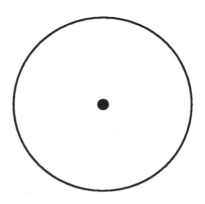

Need: Activity Master 2

Cards and Cubes

Name_____

 **E**xplore   Put dots on the number cube:
- 2 white dots for AL
- 4 red dots for NL

Your Turn

Experiment with your number cube.

The Experiment

- **Roll the number cube and record the color.**
- **Repeat 11 times.**

Tally Sheet

Red	White

Talk About It

How do your results compare with:
- those of your classmates?
- the Class Graph in *Why NL?*
- the spinner in *Match the Cards?*
- the cards in *NL First?*

Think and Tell

What makes the cards, the spinner, and the number die alike?

Need: Number cube; supply of red and white gummed dots; graph from Why NL? (page 16).

Data, Chance, and Probability Activity Book, 4-6
© 1993 Learning Resources, Inc.

Rate a Team

Name_____

xplore   Make a list of your favorite baseball teams.

Choose one team.

Rate the team by putting an X on the scale.

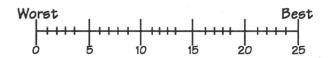

Worst   Best

0   5   10   15   20   25

Class Line Plot

• Work as a class.

• Make a large floor line plot to use as a rating scale.

• Write your rating on a paper square.

• Place your square near the number on the floor line plot.

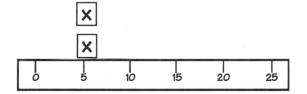

Write about the line plot:

• What is the range of ratings?

• What is the mode?

• What is the median?

How do you think the line plot would change if you rated another baseball team?

Need: Mural paper for rating scale; paper squares.

Data, Chance, and Probability Activity Book, 4-6
© 1993 Learning Resources, Inc.

Which Card?

Name_____

 **E**xplore

One Special
Baseball Card

A team or player
card from the NL
or AL

Which card do you think you will pick?

Data Activity

Study the tree diagram.

Outcomes

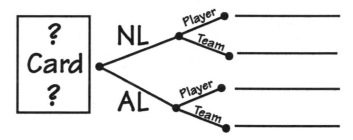

 *Talk About It*

How many different cards are possible?

What is the possibility that you will choose:

• an AL player card?

• an NL team card?

 *Think and Tell*

Make cards to show the outcomes. Can you
add these teams to the diagram?

NL: Cubs and Mets

AL: Twins and Yankees

Need: NL and AL cards (Activity
Master 2).

AL for Sure

Name_____

 **E**xplore

The packer made a mistake and put 2 AL cards in a box.

Which league do you think you will find first?

Which league do you think you will find second?

Your Turn

Find a partner.

Experiment with red/white chips:

- white side for AL
- red side for NL

The Experiment

- Put two chips under a plate, white side up.
- Draw one chip from under the plate and tell the color.
- Draw the second chip and tell the color.
- Partner takes a turn.

 Talk About It

Which league:

- is certain to be found?
- is impossible to be found?

Word Box

Impossible: an event that will not occur

 Think and Tell

On the scale write these letters:

- *A* for "Getting AL"   • *N* for "Getting NL"

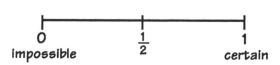

0	$\frac{1}{2}$	1
impossible		certain

Need: 2 red/white chips; paper plate.

Same Chance

Name_____

 **E**xplore

Which league do you think you will find first?

Your Turn

Find a partner.

Experiment with chips:
- 3 white for AL
- 3 red for NL

The Experiment

- Put six chips under a plate.
- Predict which color chip you'll pick.
- Draw one chip, record the color, and put the chip back.
- Partner takes a turn.
- Repeat 5 times.

Did the colors turn out as you predicted?

How do your results compare with those of other classmates?

Tally Sheet

Red	White

What fraction of the time was the chip:
- red?
- white?

Need: 3 white chips, 3 red chips.

Save: Tally Sheet for *Why the Same?* (page 23).

 22

Why the Same?

Name_____

 **xplore**   Work in small groups.

Combine the results from *Same Chance* to develop group totals.

Graph Activity

Combine the group totals from *Same Chance,* and use tally marks to make a class graph.

Total the tally marks and record below.

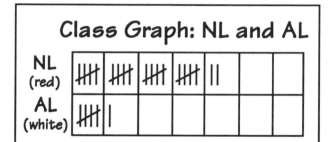

Class Graph: NL and AL

NL (red)	⌗⌗⌗	⌗⌗⌗	⌗⌗⌗	⌗⌗⌗	II		
AL (white)	⌗⌗⌗ I						

• NL _____

• AL _____

 **Talk About It**

How do the class graphs for NL and AL compare?

What are the chances of finding:

• AL first?

• NL first?

 **Think and Tell**

In *Same Chance*, there were 6 cards — 3 AL and 3 NL. On the scale show the chances of getting:

• AL   • NL

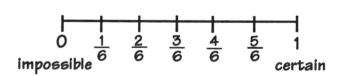

$$0 \quad \frac{1}{6} \quad \frac{2}{6} \quad \frac{3}{6} \quad \frac{4}{6} \quad \frac{5}{6} \quad 1$$

impossible   certain

Need: Data from *Same Chance* (page 22); transparency for class graph.

What Chance AL?

Name_____

 **Explore**

Free Baseball Cards

2 NL   6 AL

Which league will you find first?

Your Turn

Find a partner.

Experiment with chips:

- 6 white for AL
- 2 red for NL

The Experiment

- Put 8 chips under a plate.
- Predict the color. Draw one chip.
- Record the color on the Tally Sheet, and put the chip back.
- Partner takes a turn.
- Repeat experiment 6 times.

 **Talk About It**

Did the colors turn out as you predicted?

Tally Sheet

Red	White

 **Think and Tell**

What fraction of the time was the chip:

- red?
- white?

Need: 6 white chips, 2 red chips; paper plate.

Save: Tally Sheet for Why AL? (page 25).

 24

Why AL?

Name_____

 **Explore**   Work in small groups.

Combine the results from *What Chance AL?* to develop group totals.

Graph Activity

• Combine the group totals of *What Chance AL?*

• Write the results in the class graph with 5 tallies to a block.

• Total the tally marks, and record below.

Class Graph: NL and AL						
NL (red)	|||| |||||					
AL (white)	||||	||||	||||				

• NL _____   • AL _____

 **Talk About It**

How many tally blocks are full?

What fraction of these are AL?

What does this say about the chances of finding AL first?

 **Think and Tell**

In *What Chance AL?*, there were 8 cards — 6 AL and 2 NL. On the scale, show the chances of getting:

• AL   • NL

$$0 \quad \frac{1}{6} \quad \frac{2}{6} \quad \frac{3}{6} \quad \frac{4}{6} \quad \frac{5}{6} \quad 1$$

impossible   **certain**

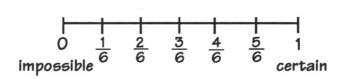
Need: Data from *What Chance AL?* (page 24).

Data, Chance, and Probability Activity Book, 4-6
© 1993 Learning Resources, Inc.

Card Match

Name_____

xplore   *What Chance AL?* had 8 free baseball cards — 6 AL and 2 NL.

Which spinners work like the cards in *What Chance AL?*

Label each shaded part *NL* and each unshaded part *AL*.

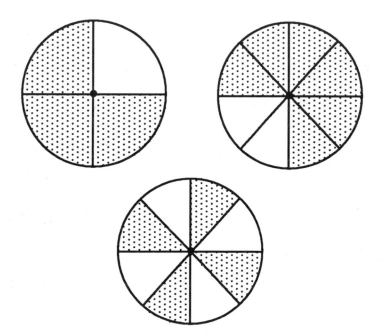

   Tell why the spinners you chose are like the baseball cards.

   Can you make another spinner that would work?

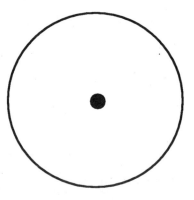

Need: Activity Master 2

Change Chances

Name _____

 **E**xplore

Free Baseball Cards

2 NL   1 AL

How would you make the chance of finding AL first the same as the chance of finding NL first?

Your Turn

- Use red and white chips to show your answer to the question above. Use red chips for NL, white chips for AL.

- Find a partner.

- Then experiment with the chips.

The Experiment

- Put your chips under the plate.
- Draw one chip, record the color on the Tally Sheet, and replace it.
- Repeat 15 times.
- Partner takes a turn.

Talk About It

In your new set of cards, is the chance for red the same as the chance for white?

Think and Tell

On the scale write the fraction that shows the chances for getting:

- red   - white

Tally Sheet

Red	White

0   1
impossible   certain

Need: 2 red chips, 2 white chips; paper plate.

Teaching Notes
Get the Message

Materials Needed:
• 2" x 2" Paper Squares • Red, Blue, and Yellow Chalk • 8-1/2" x 11" Paper

Warm-Up

The six activities in this section focus on collecting and presenting data. Students have an opportunity to explore the three kinds of data presentations: box-and-whisker plots, line plots, and stem-and-leaf plots. Approximately one week before undertaking these activities, ask students how many TV shows each one watches in a week.

Using the Pages

How Many Shows? *(page 30)*: In this activity, students work as a class to determine the *median*, *lower quartile*, and *upper quartile* for the distribution of TV shows watched in one week.

Box the TV Data *(page 31)*: When identifying the median student, advise the class to be careful when the number of students is even. In this case, the median value is the arithmetic mean of the two middle scores. As a follow-up discussion to "Write About It," ask the following questions:

> • **Is the number of data points the same on either side of the median? (Yes)**

> • **Will this always happen? (Yes)**

Note: This is an informal introduction to box-and-whisker plots. For this reason, there is no accompanying scale to the plot. Data points are merely rank ordered.

Plot the TV Data *(Page 32):* In constructing the line plot, students who watch the same number of TV shows place their squares above each other, as shown in the diagram.

Data, Chance, and Probability Activity Book, 4-6
© 1993 Learning Resources, Inc.

Make a Box-and-Whisker Plot (*page 33*): This activity introduces the typical number line scale for a box-and-whisker plot. In the "Think and Tell" section, there is a 50% probability that the new student's data point will fall within the box and a 50% chance that it will fall outside the box (25% in each whisker).

TV Time (*page 34*): A stem-and-leaf plot simplifies ordering multi-digit numbers. When students have completed the plot, they can find the range and median quickly.

TV Ratings (*Page 35*): This activity introduces the mean of a set of data and reviews the concepts of *range* and *median*. Work through the formulas with students to be sure they understand the calculations.

 **Wrap-Up**

With the type of data considered in this section, students might find points that fall on the extremes, far away from the lower and upper quartiles. As an extension of the TV viewing activity, ask students to plot the relationship of time spent on other activities to the number of TV shows viewed.

How Many Shows?

Name_____

 **Explore** About how many TV shows do your classmates watch each week?

Your Turn

- Make a list of TV shows *you* watch each week.

- Count the number of shows you watch, and write the number on your paper square.

- Work as a class.

- Tape your squares to the chalkboard in order from the fewest to the greatest number of shows.

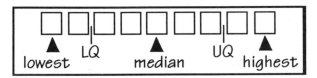

- Find the *median, upper quartile,* and *lower quartile.*

What fraction of the data is below:

- the lower quartile?
- the median?
- the upper quartile?

If there were one more student in the class, would the median change?

Word Box

Median:
the middle score

Lower Quartile (LQ):
the middle score on the top half

Upper Quartile (UQ):
the middle score on the lower half

Need: Paper squares for recording data.

Save: Paper squares for Box the TV Data (page 31).

Data, Chance, and Probability Activity Book, 4-6
© 1993 Learning Resources, Inc.

Box the TV Data

Name_____

 **xplore** What was the greatest number of TV shows watched by a student?

What was the least number?

Your Turn

- As a class, work with the squares on the chalkboard.
- Draw the vertical median line.
- Draw the vertical lines for the upper and lower quartiles.
- Complete the box.
- Place a dot at the lowest and the highest scores.
- Complete the whiskers.

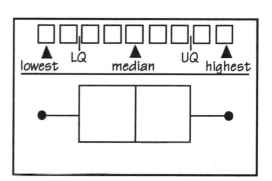

What does the box-and-whisker plot show?

How many shows per week does the typical student watch?

Need: Paper squares from *How Many Shows?* (page 30).

Plot the TV Data

Name_____

 **xplore** Mark the scale below to show how many TV shows you watch each week.

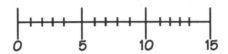

Your Turn

- Draw a large scale on the chalkboard.

- Draw a square above the scale for each classmate.

- Find the median. Color the square red.

- Find the upper and lower quartiles. Color them blue.

- Color the lowest and highest squares yellow.

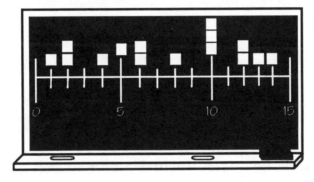

What does the line plot show?

How is the line plot like the box-and-whisker plot?

Need: Red, blue and yellow chalk.

Save: Line plot for Make a Box-and-Whisker Plot (page 33).

Data, Chance, and Probability Activity Book, 4-6
© 1993 Learning Resources, Inc.

Make a Box-and-Whisker Plot

Name_____

 **E**xplore   Where do you fit — in the box or on the whisker — of the box-and-whisker plot?

Your Turn

- Work as a class.

- Use the line plot from *Plot the TV Data* to construct a box-and-whisker plot.

- Mark the lowest and highest points.

- Draw the vertical lower quartile (LQ) and upper quartile (UQ) lines.

- Draw the vertical median (M) line, and complete the box-and-whisker plot.

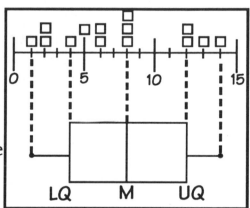

 **Talk About It**   Which fraction of the scores lay within the box?

 **Think and Tell**   If a new student arrives, what is the chance that the student's data will fall within the box?

What is the chance for the data to fall outside the box?

Need: Line plot from *Plot the TV Data* (page 32).

TV Time

Name_____

 **Explore** How many TV shows will you watch next week?

Write your estimate on a piece of paper.

Stem-and-Leaf Plot

- Make a stem-and-leaf plot on the floor as shown at right.

- Look at your number. If the number has 2 digits, fold the ten's digit under.

- Stand by the correct ten's "stem." If you have no ten's, stand by the zero "stem."

- Work with other "leaves" in your row to order yourselves from low to high, depending on your number of ones.

Number of TV Shows

Stem	Leaf
0	
1	
2	

 **Talk About It**

What is the range in number of shows?

What is the mode? the median?

 **Think and Tell**

What does the stem-and-leaf plot tell you about the number of TV shows watched?

Need: 8-1/2" x 11" paper.

 (34)

Data, Chance, and Probability Activity Book, 4-6
© 1993 Learning Resources, Inc.

TV Ratings

Name_____

 **xplore**   What shows do you like?

Make a list of 10 TV shows, and number them 1 to 10.

Rank each show on the scale below. Use each rating only once.

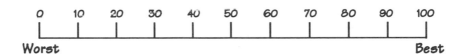

Data Activity

- Make a chart of your ratings in order from highest to lowest.

- Find the range, median, and mean.

TV Ratings Chart	
StarTime	100
Roller Hit	85
Big Family	80
Jet Set	70
Cartoon-O's	65
The Twins	55
Fortune	50
News	40
Sports	35
Rescue	20
Total ratings	____

Talk About It   How do you find the range, median, and mean?

Think and Tell   Compare your ratings with classmates. How are your ratings alike?

How are your ratings different?

Word Box

Mean: total ratings divided by number of data

Need: Chart paper.

Data, Chance, and Probability Activity Book, 4-6
© 1993 Learning Resources, Inc.

Teaching Notes
Candy Sort

Materials Needed:

• Number Cubes • Chart Paper • 2" x 2" Paper Squares • Red and White Chips • Paper Plates

Warm-Up

This section investigates numerical probabilities for various situations. To introduce this series of activities, place a sample of 50 marbles (5 each of 10 colors) in a bag. Have students predict the most likely color when a single marble is drawn, then check their predictions.

Using the Pages

Take a Candy *(page 38)*: Help students conclude that the results of the number cube trials provide the answer to the candy problem.

Candy Chances *(page 39)*: By combining the individual outcomes in the class graph, results should approach a 2/6 ratio.

Match the Candies *(page 40)*: Ask students to explain their choices of spinners. Point out that the spinner divided into thirds could also be used to obtain the answer, since there are three colors of candy, each represented by one space on the spinner.

Mini Candies *(page 41)*: Since the chips are drawn *without replacement*, the probability of two red chips on the end is 1/6. Note that in the second draw, the probability of red is 1/3 if 1 red and 2 whites are left, or 2/3 if 2 reds and 1 white are left.

Double Red *(page 42)*: When individual results are combined in the class graph, the outcome should approach a 1/6 ratio.

Mini Packs *(page 43)*: If only red chips are added, three are needed so that each color will have the same chance of being drawn first.

More Mini Packs *(page 44)*: The chance of getting white first is 3/4 or any equivalent fraction. To make the chances the same for each color candy, there are several ways of proceeding: remove 2 whites (1 white, 1 red); add 2 reds (3 reds, 3 whites); or produce any arrangement with an equal number of both colors.

Data, Chance, and Probability Activity Book, 4-6
© 1993 Learning Resources, Inc.

Better Chances *(page 45)*: After students complete the experiments, help them conclude that the most likely color is the color of the candy that exists in greater quantity.

Greater Chance *(page 46)*: Accept any arrangement with more red chips than white.

Wrap-Up

In *Mini Candies*, students estimated the chance of having 2 reds first in the roll. (This turned out to be 1/6, even though the chance of a single red was 1/2.) To further explore the probability of repeating an event twice, ask the following question: *Is the probability that a basketball player "sinks" a single free throw higher or lower than making 2 free throws in a row?* Students can experiment in physical education class or at home to generate data for answering this question.

Take a Candy

Name_____

 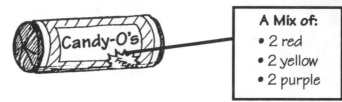

Candy-O's

A Mix of:
- 2 red
- 2 yellow
- 2 purple

When you open the candies, which color will you find first?

Your Turn

Experiment with a
number cube:

- 1,2 - red
- 3,4 - yellow
- 5,6 - purple

The Experiment

- Predict the outcome if you roll the number cube once.
- Roll the number cube and record the color on the Tally Sheet.

Why does the number cube work like the candies?

Did the colors turn out as you predicted?

Tally Sheet

Red	Yellow	Purple

What fraction of the time will the first candy be:

- red?
- yellow?
- purple?

Need: 1 number cube

Save: Tally Sheet for Candy Chances (page 39).

Data, Chance, and Probability Activity Book, 4-6
© 1993 Learning Resources, Inc.

Candy Chances

Name _____

Explore   Work in a small group.

Combine the results from *Take a Candy* to develop group totals.

Graph Activity

- Work as a class to combine the results.

- Use 1 colored square for each tally block of 5.

- Tape the squares on chart paper to make a class graph.

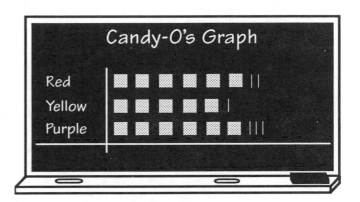

Candy-O's Graph

Red	
Yellow	
Purple	

How many squares are there in total?

What fraction of candies are the following colors:

    • red?   • yellow?   • purple?

What are the chances for each color being first in the roll?

Think and Tell

On the scale, show the chances of each color being first:

    • red (R)   • yellow (Y)   • purple (P)

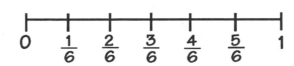

$$0 \quad \frac{1}{6} \quad \frac{2}{6} \quad \frac{3}{6} \quad \frac{4}{6} \quad \frac{5}{6} \quad 1$$

Need: Chart paper; 2" x 2" paper squares: red, yellow, and purple data from *Take a Candy* (page 38).

Match the Candies

Name_____

 **Explore**

Take a Candy had 6 candies — 2 red, 2 yellow, and 2 purple.

Which spinners work like the roll of candies?

Label the 3 colors in the spinners.

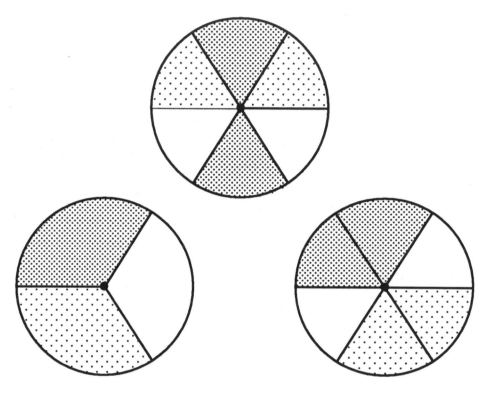

 **Talk About It**

Tell why the spinners you chose are like the roll of Candy-O's.

 **Think and Tell**

How are the three spinners alike? different?

Mini Candies

Name _____

 **Explore**

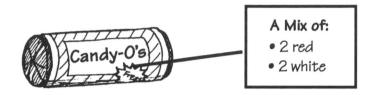

Candy-O's

A Mix of:
• 2 red
• 2 white

Estimate the chances of 2 red candies being first in the roll.
(Circle one.)

1 of 6   3 of 6

2 of 6   4 of 6

Your Turn

Work with a partner.

Experiment with chips:
• 2 red (R)
• 2 white (W)

The Experiment

• Put 4 chips under a plate.
• Draw one chip. Do not put it back.
• Draw another chip.
• Record the colors on the Tally Sheet.
• Partner takes a turn.

Talk About It

What fraction of the tallies were RR?

Tally Sheet

RR	Not RR

Think and Tell

How does the fraction for RR compare with your estimate?

Need: 2 red chips, 2 white chips.

Save: Data for *Double Red* (page 42).

Data, Chance, and Probability Activity Book, 4-6
© 1993 Learning Resources, Inc.

Double Red

Name _____

 **Explore**   Work in small groups.

Combine the results from *Mini Candies* to develop group totals.

Graph Activity

- Work as a class to combine the results. Use 1 square for each tally block of 5.

- Tape the squares on chart paper to make a class graph.

Mini Candy-O's Class Graph

RR ▨▨▨ ||

Not RR ▨▨▨▨▨▨▨▨▨▨▨ |||

 **Talk About It**

How many squares are there in all?

About what fraction of these are RR?

What does this say about the chances that the first two candies in the roll will be red?

 **Think and Tell**

On the scale, show the chances of having two reds first.

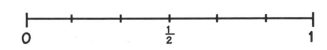

0          $\frac{1}{2}$ 1

Need: 2" x 2" paper squares; chart paper; data from *Mini Candies* (page 41).

 42

Mini Packs

Name_____

 **E**xplore

 Candy-O's

A Mix of:
• 4 white
• 1 red

How many red candies need to be added so that each color has the same chance of being first in the roll?

Your Turn

• Use red and white chips to show your answer.

• Then experiment with the chips.

The Experiment

• **Put the chips under a plate.**
• **Draw one chip, tally the color, and put the chip back under the plate.**
• **Repeat 20 times.**

Tally Sheet

Red	White

 Talk About It

Did your tally match your prediction?

Why or why not?

 Think and Tell

Write a fraction to show the chance of red first —

• *before* you added candies   • *after* you added candies

0 ⊢———————————⊣ 1

Need: 5 red chips, 5 white chips; paper plate.

Data, Chance, and Probability Activity Book, 4-6
© 1993 Learning Resources, Inc.

More Mini Packs

Name_____

 **E**xplore

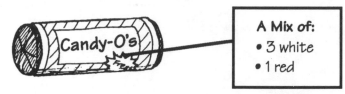

Candy-O's

A Mix of:
- 3 white
- 1 red

Write the fraction that tells the chance of a white candy being first in the roll.

How could you even the chances of getting a red or white candy?

Your Turn

- Use red and white chips to show your answer to the second question above.

- Then experiment with the chips.

The Experiment

- **Put the chips under a plate.**
- **Draw one chip, record the color, and put the chip back under the plate.**
- **Repeat 20 times.**

Tally Sheet

Red	White

Talk About It

Did your tally match your prediction? Why or why not?

Think and Tell

Did everyone solve the problem the same way?

Need: 5 red chips, 5 white chips; paper plate.

 44

Better Chances

Name _____

 **E**xplore

 Candy-O's

A Mix of:
• 3 white
• 1 red

Which color has a better chance of being first in the roll?

If 3 red candies are added, what color has a better chance?

Your Turn

Partner's Experiment

• Use red and white chips to model the first pack.
• Place the chips under a plate.
• Draw one chip, record the color, and put the chip back under the plate.
• Repeat 20 times.

Your Experiment

• Use red and white chips to model the second pack.
• Place the chips under a plate.
• Draw one chip, record the color, and put the chip back under the plate.
• Repeat 20 times.

 **Talk About It**

How do your results compare?

Do they match your predictions?

Tally Sheet

Red	White

Tally Sheet

Red	White

 **Think and Tell**

Write the fraction that shows the chance of "red first":

• before adding candies   • after adding candies

0 ———————————————— 1

Need: 8 red chips, 8 white chips; 2 paper plates.

Greater Chance

Name_____

 xplore

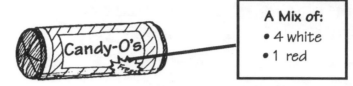

A Mix of:
- 4 white
- 1 red

Write the fraction that tells the chance of a white candy being first in the roll.

How could you make the chance of getting a red candy first greater than the chance of getting a white candy first?

Your Turn

- Use red and white chips to show your answer to the second question above.

- Then experiment with chips.

The Experiment

- Put the red and white chips under a plate.
- Draw one chip, record the color, and put the chip back under the plate.
- Repeat 20 times.

Tally Sheet

Red	White

Talk About It

Did your tally match your prediction? Why or why not?

Think and Tell

On the scale below write the fraction that shows the chance of a red candy first after you changed the number of candies.

```
0                    1
```

Need: 5 red chips, 5 white chips, paper plate.

Data, Chance, and Probability Activity Book, 4-6
© 1993 Learning Resources, Inc.

Teaching Notes
Pizza Toss

Materials Needed:
• Red and White Chips • Gummed Labels •
8- 1/2" x 11" Paper • Chart Paper •
Chalkboard • Colored Pencils or Markers

Warm-Up

This section presents several data and chance experiences involving pizza. To introduce this section, ask the class to predict how many different pizzas can be ordered at a favorite pizza place. Discuss the possible sizes of pizzas, numbers of toppings, types of crust, and so on.

Using the Pages

Kinds of Pizza *(page 49)*: More chips are provided than needed, so you may need to monitor the way students set up the chips. This activity applies the multiplication principle: the number of different kinds of pizza available from the menu is equal to 2 (crusts) x 3 (toppings).

Pizza Pieces *(page 50)*: Review the construction of a stem-and-leaf plot. Help students conclude that this plot helps them identify each quartile, the range and median.

Box the Pizza *(page 51)*: Review the information on the line plot to construct the box-and-whisker plot. Using the example shown, point out that the data are slightly skewed towards the third quartile.

Pizza Favorites *(page 52)*: Suggest that students use their sample data to predict outcomes for larger or smaller populations.

Topping Survey *(page 53)*: Work through the example with students to show how a proportion can be used to apply the results to larger groups.

Pizza Data *(page 54)*: Ask students to collect the data. Remind them to obtain responses from similar-sized age groups, so they get valid results. After all data are collected, help students create a chart.

Pizza Show (*page 55*): This activity lets students display their results in a variety of formats including bar graph and line graph. Encourage students to discuss the format they find represents the data most appropriately. Challenge older children to make circle graphs of the data.

Wrap-Up

As an extension of the *Pizza Favorites* activity, the class might consider a list of unusual topping choices that would give completely different results. Encourage the class to find examples of line plots and stem-and-leaf plots in their textbooks, newspapers, or magazines. Ask them to think about the ways people use plots like these to make decisions about the production and marketing of food.

Kinds of Pizza

Name_____

 **Explore**

How many different kinds of pizza can you order?

Your Turn

Use chips:
  • red (R) for thick   • white (W) for thin

Mark the chips:
  • 1 sausage   • 2 cheese   • 3 pepperoni

Show the different kinds of pizza you can order. List them.

> R1

How many different kinds of pizza did you find?

What if another ingredient, mushrooms, were added to the menu? How many different kinds of pizza could you order?

> Need: 4 red chips, 4 white chips; gummed labels for chips.

Data, Chance, and Probability Activity Book, 4-6
© 1993 Learning Resources, Inc.

Pizza Pieces

Name_____

 **E**xplore Estimate how many pizza pieces your family eats in a week.

Write the number on a piece of paper.

Floor Stem-and-Leaf Plot

- Make a stem-and-leaf plot on the floor.

- Look at your number. If the number has 2 digits, fold the ten's digit under.

- Stand by the correct ten's "stem." If you have no ten's, stand by the zero "stem."

- Work with other "leaves" in your row to order yourselves from low to high, depending on your number of ones.

- Place the numbers on the floor. Copy the stem-and-leaf plot on chart paper.

Stem	Leaf
0	
1	
2	

 What does the stem-and-leaf plot tell you about pizza-eating habits?

 How could you use the stem-and-leaf plot to find:
- the median?
- the lower quartile?
- the upper quartile?

Data, Chance, and Probability Activity Book, 4-6
© 1993 Learning Resources, Inc.

Box the Pizza

Name_____

 **Explore**   Use the Stem-and-Leaf Plot from *Pizza Pieces* to find:

- the lowest number   • the highest number   • the median
- the lowest quartile   • the upper quartile

Your Turn

- Make a number line on the chalkboard. Mark the lowest and highest numbers.

- Draw the vertical LQ and UQ lines.

- Draw the vertical M line to complete the box-and-whisker plot.

Box-and-Whisker Plot

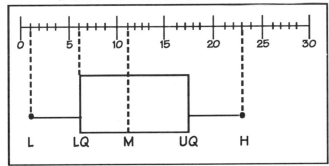

 What fraction of the numbers lie:

- within the box?   • within the whiskers?

 Where did your number lie on the box-and-whisker plot? Tell what this means.

Need: Stem-and-leaf plot from *Pizza Pieces* (page 50); chalkboard.

Pizza Favorites

Name_____

 **E**xplore   What is your favorite pizza topping?

Data Activity

- Look at the data at right. 100 people were asked to identify their favorite pizza topping.

- What is the probability that people would choose pepperoni?

- To find the probability, total the number of students choosing pepperoni and divide that number by 100.

Pizza Choices	
Topping	**Frequency**
Sausage	26
Cheese	29
Pepperoni	20
Onions	8
Mushrooms	10
Green Peppers	7
Total	100

 **Talk About It**   What is the probability of persons' choosing:

- sausage?   • mushrooms?
- cheese?   • onions?

Word Box

Probability: ratio of the number of successes to the total possibilities

 **Think and Tell**   If you were to survey 200 people, how many people would you expect to choose pepperoni?

Data, Chance, and Probability Activity Book, 4-6
© 1993 Learning Resources, Inc.

Topping Survey

Name_____

 **E**xplore   What's the favorite pizza topping in your class?

Data Activity

- As your teacher polls the class, make a chart like that shown below.

- Use the information to predict the expected frequency for a group twice the size of your class.

Pizza Topping Chart

Topping	Tally	Frequency
Cheese		
Sausage		
Pepperoni		

 Describe how you determined the expected frequency.

 How would you make predictions for a group 3 times bigger than your class?

Need: Chart paper.

Data, Chance, and Probability Activity Book, 4-6
© 1993 Learning Resources, Inc.

Pizza Data

Name_____

xplore   Which age groups eat pizza most frequently?
Make a prediction.

Your Turn

- Take a survey in your area.

- Make a chart using the example shown at right.

- Choose one age group and ask 20 people, "Did you eat pizza last week?"

- Pool your data with your classmates and complete your chart.

Pizza Lovers		
Age Group in Years	Tally	Number who said "Yes"
5-10		
11-15		
16-20		
21-25		
26 & up		

Which group had the highest percentage of "Yes" replies?
Which had the lowest?

If you asked 200 5- to 10-year-olds the same questions, how
many would you expect to answer "Yes?"

| Need: Chart paper. |

| Save: Chart for *Pizza Show* (page 55). |

Data, Chance, and Probability Activity Book, 4-6
© 1993 Learning Resources, Inc.

Pizza Show

Name_____

 xplore How can you show the data from your pizza lovers' survey?

Graph Activity

- Look at the chart you made in *Pizza Data.*

- Think about the ways to show your data.

- Make bar and line graphs to show your data.

- A sample of one group, 5- to 10-year-olds, is shown for you.

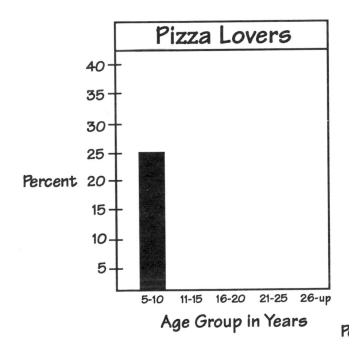

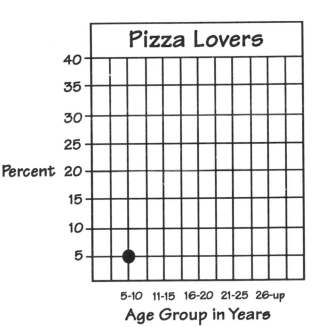

Need: Paper, colored pencils or markers, chart from *Pizza Data* (page 54).

Just as Many

Name_____

xplore

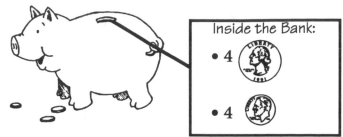

Inside the Bank:
- 4 ⓠ
- 4 ⓓ

Shake the bank. Is any coin more likely to come out first?

Your Turn

- Find a partner.
- Use Spinner B on Activity Master 1 (page 75).

The Experiment

- Take turns.
- Spin the spinner 16 times.
- Tally the results.

Tally Sheet

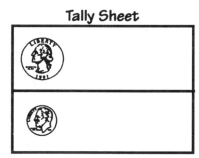

Talk About It

Did the outcome turn out as you predicted?

Why?

How do your results compare with those of other classmates?

Why does Spinner B model the bank?

Think and Tell

On the scale show the chances of getting:

- a 25¢ coin
- a 10¢ coin

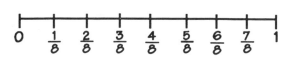

$$0 \quad \frac{1}{8} \quad \frac{2}{8} \quad \frac{3}{8} \quad \frac{4}{8} \quad \frac{5}{8} \quad \frac{6}{8} \quad \frac{7}{8} \quad 1$$

Need: Activity Master 1 (page 75), paper clip.

Data, Chance, and Probability Activity Book, 4-6
© 1993 Learning Resources, Inc.

3 Coins

 **xplore**

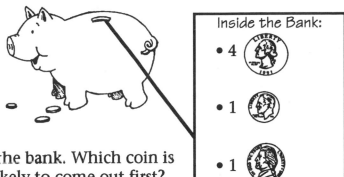

Inside the Bank:
• 4
• 1
• 1

Shake the bank. Which coin is most likely to come out first?

Your Turn

• Find a partner.
• Use Spinner C on Activity Master 1 (page 75).

The Experiment

• Take turns.
• Spin the spinner 24 times.
• Tally the results.

Tally Sheet

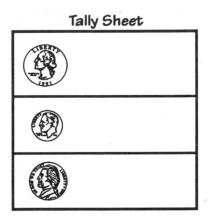

 **Talk About It**

Did the outcome turn out as you predicted? Why?

How do your results compare with those of other classmates?

Why does the spinner model the 3-coin bank?

 **Think and Tell**

On the scale show the chances of getting:

• a 5¢ coin   • a 10¢ coin   • a 25¢ coin

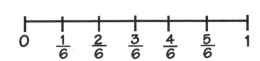

$$0 \quad \frac{1}{6} \quad \frac{2}{6} \quad \frac{3}{6} \quad \frac{4}{6} \quad \frac{5}{6} \quad 1$$

Need: Activity Master 1 (page 75), paper clip.

Data, Chance, and Probability Activity Book, 4-6
© 1993 Learning Resources, Inc.

 59

Even It Up

Name_____

 xplore

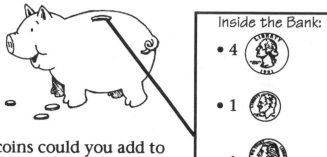

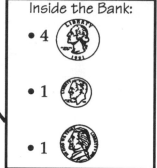

Inside the Bank:

• 4 [quarter]

• 1 [dime]

• 1 [nickel]

What coins could you add to the bank to give all coins the same chance of being first?

Your Turn

- Find a partner.
- Set up your new bank (bag) so that all coins have the same chance of coming out first.

The Experiment

- Draw a coin, tally the result, and put the coin back in the bag.
- Repeat 15 times.
- Partner takes a turn.

Tally Sheet

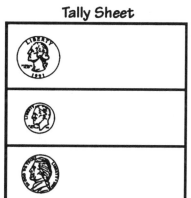

Talk About It

Is your new bank fair to all coins?

How do your results compare with those of other classmates?

Think and Tell

On the scale show the chances of getting:

- a 5¢ coin   • a 10¢ coin   • a 25¢ coin

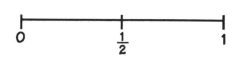

0          $\frac{1}{2}$ 1

Need: 5 each of 25¢, 10¢, and 5¢ coins; bag.

Data, Chance, and Probability Activity Book, 4-6
© 1993 Learning Resources, Inc.

Nickel Up

Name_____

 **Explore**

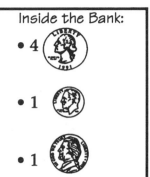

Inside the Bank:

- 4
- 1
- 1

What is the fewest number of nickels you could add to give the nickel the best chance of coming out first?

Your Turn

- Find a partner.
- Set up your new bank (bag) so the nickel has the best chance.

Tally Sheet

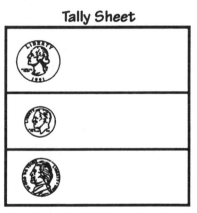

The Experiment

- Draw a coin, record, and put the coin back in the bag.
- Repeat 15 times.
- Partner takes a turn.

Talk About It

Does the outcome match your predictions? Why or why not?

Think and Tell

On the scale write the fraction that shows the chance of getting:

- a quarter   • a dime   • a nickel

0 ———————————— 1

Need: 1 nickel, 1 dime, and 4 quarters; bag.

Data, Chance, and Probability Activity Book, 4-6
© 1993 Learning Resources, Inc.

Two Come Out

 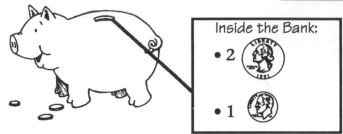

Inside the Bank:
- 2 [quarter coin]
- 1 [dime coin]

Two coins come out, one after the other.

How many different ways could the coins come out?

Your Turn

Experiment with the coins. Show the ways. List them.

How many different ways did you find?

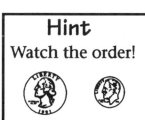

Hint
Watch the order!

is different from

How does order make a difference in counting the ways?

A Chance for Two

Name_____

 **E**xplore

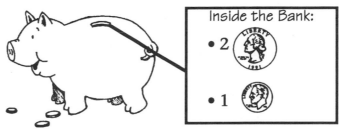

Inside the Bank:
- 2
- 1

Two coins come out, one after the other. Which is more likely:

- two the same?   •two different?

Your Turn

Find a partner. Put the coins in the bag.

The Experiment

- Shake the bag.
- Draw 1 coin. (Do not replace it.)
- Draw a second coin.
- Tally the outcomes.
- Repeat 10 times.
- Partner takes a turn.

Tally Sheet

1st coin	2nd coin

 Talk About It

Did the outcome turn out as you predicted?

How do your results compare with those of others?

 Think and Tell

On the scale show the chances of getting:

- two the same   • two different

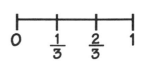

0    $\frac{1}{3}$ $\frac{2}{3}$ 1

Need: 2 quarters and 1 dime; bags; data from
Two Come Out (page 62).

Match the Coins

Name_____

xplore   Which spinners work like the bank in *A Chance for Two?*

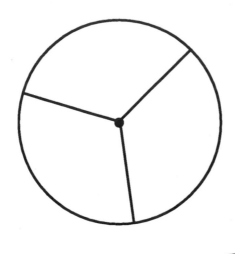

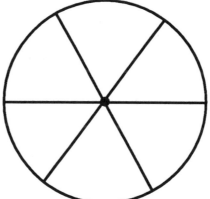

Tell why.

If another spinner would work, show it.

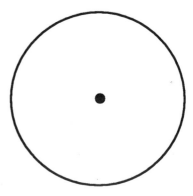

Data, Chance, and Probability Activity Book, 4-6
© 1993 Learning Resources, Inc.

Teaching Notes
Game Time

Materials Needed:
• Coins • Red/White Counters (chips) • Red
Counters (chips) • White Counters (chips)

Warm-Up

Introduce this section on games by using a number cube with 4 red and 2 white
dots. Divide the class into two groups, and play a game in which a group gets one
point every time its color comes up on the number cube. The first group to get 5
points wins the game. On the first round, let Group 1 choose the color (red or
white). On the second round, Group 2 chooses first.

Following the games, discuss:

> • **whether the game was fair**
>
> • **what could be done to make the game fair**

In this section, students will play various fair and unfair games. In these games,
the criteria for a "fair" game is that each of the players has an equal chance to win.

Using the Pages

Who Goes First? *(page 67)*: Review the definition of *fair* from the Word Box.
After students complete the page, help them conclude that tossing a coin is a fair
way to decide who takes the first turn.

Now Who? *(page 68)*: This game is fair since both Tim and Tami have two out of
four possible outcomes: TT and HH for Tim, and TH and HT for Tami. Students
may need to convince themselves that TH and HT are, in fact, two different
outcomes.

Are the Chips Fair? *(page 69)*: Students will find only an
experimental result for this activity. The game is fair since
there are two equally likely outcomes: RR and RW.

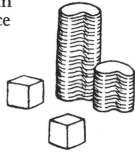

Data, Chance, and Probability Activity Book, 4-6
© 1993 Learning Resources, Inc.

Is It Fair? *(page 70)*: As students experiment with this game, they will find it is grossly unfair. Help students discover that the probability for Jillian winning is 5/6 and the probability for Peter winning is only 1/6.

 Wrap-Up

Students might be challenged to create their own "fair" games, play them with partners, then write a short paper reporting their results. Be sure to have them defend why their games are fair.

Who Goes First?

Name_____

 **Explore**   Tami and Tim toss a coin to decide who takes the first turn at a video game. Is it fair to toss a coin?

Your Turn

- Find a partner and take turns.
- Toss a coin.
- Tally the result: Heads or Tails.
- Repeat 11 times.

Work together to find the class totals and record them.

- Number of heads _____
- Number of tails _____

Tally Sheet

Heads	Tails

 **Talk About It**   How does the number of Heads and Tails compare with:

- your partner's data?
- the class data?

Word Box

Fair: all players have an equal chance

 **Think and Tell**   Is tossing a coin a fair way to decide who goes first?

Why?

Need: 1 coin.

Data, Chance, and Probability Activity Book, 4-6
© 1993 Learning Resources, Inc.

67

Now Who?

Name_____

 **xplore**   Tami and Tim toss 2 coins to decide who has the first turn at a video game. If both coins land on the same side, Tami goes first. If not, Tim goes first.

Your Turn

- Find a partner and take turns.
- Toss 2 coins.
- Record Same or Different.
- Repeat 10 times.

Tally Sheet

Same	Different

Work together to find the class totals and record them:

- Same_____

- Different_____

 **Talk About It**   How do the numbers for Same and Different compare with:

- your partner's data?
- the class data?

 **Think and Tell**   Is tossing two coins a fair way to decide?

How can this be? Tami *seems* to have 2 chances for same, and Tim only has 1 chance for different.

Need: 2 coins.

 68

Data, Chance, and Probability Activity Book, 4-6
© 1993 Learning Resources, Inc.

Are the Chips Fair?

Name_____

 **Explore**   Tami and Tim toss 2 chips to decide who has the first turn at a video game. One chip is red on both sides. The other is red on one side and white on the other.

- Tami wins if the chips match.
- Tim wins if the chips do not match.

Your Turn

- Find a partner and take turns.
- Toss the 2 chips.
- Record Match or No Match.
- Repeat 10 times.
- Work together to find the class totals and record them:

  - Match _____

  - No Match _____

Tally Sheet

Match	No Match

 How do the numbers for Match and No Match compare with:

- your partner's data?
- the class data?

 Is this a fair way to decide? Why?

Need: 1 red/white chip, 1 red chip.

Data, Chance, and Probability Activity Book, 4-6
© 1993 Learning Resources, Inc.

Is It Fair?

Name_____

 **Explore** Peter and Jillian use 2 red chips and 1 white chip in a game. Peter mixes the chips in his hands behind his back. He leaves 2 chips in one hand and 1 in the other.

Jillian picks one hand. If it has a red chip, she wins. If not, Peter wins. Is this game fair?

Your Turn

- Find a partner and take turns.
- Play the game 10 times.
- Record the results.
- Work together to find the class totals and record them:
  - Peter wins
  - Jillian wins

Tally Sheet

Peter	Jillian

 **Talk About It** How do the number of wins for Peter and Jillian compare with:
- your partner's data?
- the class data?

 **Think and Tell** Is this a fair game? Why?

Need: 2 red chips, 1 white chip.

 70

Teaching Notes
Pancake Fun

Materials Needed:
• Dice • Activity Master 2 (page 76)

Warm-Up

In this series of activities, students simulate real-world problems with random devices. To begin, ask students how they would simulate the chance that the last digit of someone's telephone number would be even. In this situation, students can toss a coin since there is a 50-50 chance of odd-even (equal chance of heads or tails).

Using the Pages

Blueberry Pancakes *(page 72)*: Discuss why the number die is a suitable model for randomly distributing the blueberries. Each of the six pancakes can be made to correspond to one of the numbers on the face of the die. To get a good estimate, suggest that students combine their results. The estimate should be around 83/100.

No Blueberries *(page 73)*: This exercise is almost identical to *Blueberry Pancakes* except that cards are used instead. The estimate should be about seven.

Enough Blueberries *(page 74)*: Students should find that each pancake has at least one blueberry, since the cards are drawn until each card appears at least once. The estimate should be around 100.

Wrap-Up

Students can explore other real life problems and determine ways to simulate them. Examples can be found in all facets of our daily lives – from sports to the weather to scientific discoveries.

PANCAKE BREAKFAST

Blueberry Pancakes

Name_____

 **E**xplore

A mix for 6 pancakes has 20 blueberries.

What is the chance of getting a pancake with at least two blueberries?

Your Turn

Number the pancakes 1 to 6 to match the numbers on a die.

Roll the die 20 times, once for each blueberry.

Tally each number rolled in the box for that number.

1	2	3	4	5	6

Talk About It

If each tally mark stands for a blueberry, how many pancakes get at least two blueberries?

Think and Tell

Based on your data, what is the probability of getting a pancake with two blueberries?

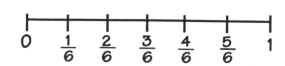

Need: One die.

No Blueberries

Name _____

 **E**xplore

A mix for 26 pancakes has 40 blueberries.

What is the chance of getting a pancake with *no* blueberries?

Your Turn

Make 26 cards, and number the cards from 1 to 26.

Mix the cards and spread them face down.

Take a card and record the number with a tally mark.

Mix the cards and repeat 40 times, one for each blueberry.

1	2	3	4	5	6	7	8	9	10	11	12	13
14	15	16	17	18	19	20	21	22	23	24	25	26

If each tally mark stands for a blueberry, how many pancakes have *no* blueberries?

Based on your data, what is the chance of getting a pancake with no blueberries?

Need: Activity Master 2.

Save: Cards for Enough Blueberries (page 74).

Enough Blueberries

Name_____

 **E**xplore

Use a mix for 26 pancakes.

How many blueberries will you need so that every pancake has at least one blueberry?

Number you predict:_____

Your Turn

- Use the set of 26 cards, numbered 1 to 26, to represent the 26 pancakes.
- Mix the cards and spread them face down.
- Take a card and record the number with a tally mark.
- Replace the card in the deck and mix.
- Repeat until each number has one tally mark.

1	2	3	4	5	6	7	8	9	10	11	12	13
14	15	16	17	18	19	20	21	22	23	24	25	26

If each tally mark stands for a blueberry, how many blueberries will you need?

Based on your data, was the number of blueberries you predicted correct?

Need: Card deck from *No Blueberries* (page 73).

Data, Chance, and Probability Activity Book, 4-6
© 1993 Learning Resources, Inc.

Activity Master 1

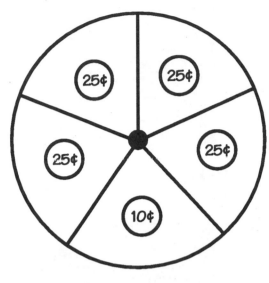

Spinner A

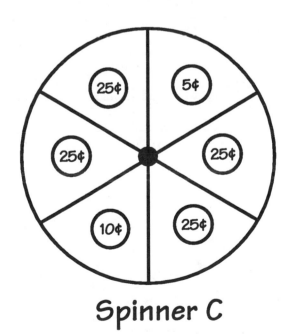

Spinner C

Spinner B

Activity Master 2
NL and AL Cards, Pancake Card

Number_____

Number_____

Data, Chance, and Probability Activity Book, 4-6
© 1993 Learning Resources, Inc.

Data, Chance & Probability

Family-Gram

Dear Family,

Congratulations! Your child

_____

has successfully completed the
unit entitled

_____

in Data, Chance, and Probability
Activity Book, 4-6.

_____ **Date** _____ **Teacher**

Data, Chance & Probability

Award Certificate

Congratulations to:

_____

for excellent work in the unit entitled

_____

in Data, Chance, and Probability Activity Book, 4-6.

Date _____ Teacher _____

Data, Chance, and Probability Activity Book, 4-6
© 1993 Learning Resources, Inc.

Data, Chance & Probability

GOOD WORK AWARD

To: _____

For: _____

Date _____

Teacher _____

rogress Chart

Name_____

Grade_____

Date Started_____

Date Finished_____

Distances

- [ ] How Far?
- [ ] How Many?
- [ ] Graph It!

Baseball Cards

- [ ] Free Baseball Card
- [ ] 3-Card Special
- [ ] NL First
- [ ] Why NL?
- [ ] Match the Cards
- [ ] Cards and Cubes
- [ ] Rate a Team
- [ ] Which Card?
- [ ] AL for Sure
- [ ] Same Chance
- [ ] Why the Same?
- [ ] What Chance AL?
- [ ] Why AL?
- [ ] Card Match
- [ ] Change Chances

Get the Message

- [ ] How Many Shows?
- [ ] Box the TV Data
- [ ] Plot the TV Data
- [ ] Make a Box-and-Whisker Plot
- [ ] TV Time
- [ ] TV Ratings

Candy Sort

- [ ] Take a Candy
- [ ] Candy Chances
- [ ] Match the Candies
- [ ] Mini Candies

- [ ] Double Red
- [ ] Mini Packs
- [ ] More Mini Packs
- [ ] Better Chances
- [ ] Greater Chance

Pizza Toss

- [ ] Kinds of Pizza
- [ ] Pizza Pieces
- [ ] Box the Pizza
- [ ] Pizza Favorites
- [ ] Topping Survey
- [ ] Pizza Data
- [ ] Pizza Show

Out of the Bank

- [ ] In the Bank
- [ ] Just as Many
- [ ] 3 Coins
- [ ] Even it Up
- [ ] Nickel Up
- [ ] Two Come Out
- [ ] A Chance for Two
- [ ] Match the Coins

Game Time

- [ ] Who Goes First?
- [ ] Now Who?
- [ ] Are the Chips Fair?
- [ ] Is It Fair?

Pancake Fun

- [ ] Blueberry Pancakes
- [ ] No Blueberries
- [ ] Enough Blueberries

Data, Chance, and Probability Activity Book, 4-6
© 1993 Learning Resources, Inc.